Yao Jeannot Kouakou
Mélanie Blanchard

Agro-ecology or organic farming in Sub-Saharan Africa

Yao Jeannot Kouakou
Mélanie Blanchard

Agro-ecology or organic farming in Sub-Saharan Africa

ScienciaScripts

Cover image: www.ingimage.com

This book is a translation from the original published under ISBN 978-613-8-42441-3.

Publisher:
Sciencia Scripts
is a trademark of
Dodo Books Indian Ocean Ltd. and OmniScriptum S.R.L publishing group

120 High Road, East Finchley, London, N2 9ED, United Kingdom
Str. Armeneasca 28/1, office 1, Chisinau MD-2012, Republic of Moldova, Europe
Printed at: see last page
ISBN: 978-620-7-30076-1

TABLE OF CONTENTS:

Summary

The way biomass is managed on mixed crop-livestock farms is very important for the production of a sufficient quantity of organic manure. In organic farming, producers have no alternative but to produce and apply organic manure to ensure the sustainability and productivity of their farming systems, as mineral fertilisers are not authorised and other alternative solutions are not accessible. We therefore hypothesise that farmers involved in organic farming have innovative biomass management practices that could serve as a model for the development of sustainable agriculture. In south-west Burkina Faso, we identified biomass management practices among a sample of 30 heads of farms engaged in organic farming, using a systemic analysis. The surveys revealed different biomass management methods for different types of producer. Agri-breeders and livestock farmers, being the best equipped in terms of agricultural and transport equipment, are those who optimise biomass management the most and have the best harvest yields. Farmers are characterised by their modest operations, limited transport equipment and the lowest yields. On the whole, they are all faced with problems linked to the climatic situation and difficulties in transporting organic manure due to a lack of transport equipment.

Key words: biomass management / food / mixed farming / organic farming, organic manure.

Introduction

The Green Revolution substantially increased agricultural production and enhanced food security worldwide (FAO 2014). But in many countries, intensive farming based on this production model has depleted natural resources, compromising future productivity (Blanchard *et al.,* 2006). It is a production model that relies heavily on inputs: water, land, energy, fertilisers, insecticides and herbicides, and technologies. In order to cope with these environmental constraints and to meet the increased demand for food products, maintaining soil fertility remains a major condition for improving agricultural productivity and sustainability (Bationo *et al.*, 2007; Sedogo, 1981). Studies carried out by (Blanchard *et al.,* 2016); (Vall *et al.,* 2016); (Griffon 2009) have shown that ecological intensification would appear to be the preferred change and would be more attractive to adopt, taking into account environmental issues (climatic hazards) and sustainability (soil fertility). It is in this context that agriculture-livestock integration takes on its importance and is seen as an opportunity to participate in this intensification, and seems to be a strategy for improving producers' performance. However, it is a practice that requires twofold investment, particularly in animal feed and the production of organic manure. In western Burkina Faso, new production systems are emerging in response to these principles and market opportunities. Some farms are embarking on these two paths to ecological intensification, the first of which is linked to conversion to organic farming via the introduction and development of organic cotton in the region, and the second to adjustments to production systems on the most flexible farms, when limitations are encountered (soil infertility, lack of money, lack of labour, etc.). But the disadvantage is that biomass management on farms in the zone is not optimal. Farmers in the area are characterised by what is known as atypical management of organic manure (Tingueri, 2015). The hypothesis of this study is that mixed crop-livestock farms involved in agro-ecology or organic farming implement innovative biomass management practices (crop residues, fodder, effluents) that are efficient in terms of productivity and environmental protection. These practices can be used to develop productive, sustainable agriculture in western Burkina Faso as a source of innovation for mixed crop-livestock farms not involved in specific commodity chains.

Against this background, we set ourselves the general objective of identifying efficient biomass management practices on mixed crop-livestock farms involved in agro-ecological or organic farming systems. Specifically, the aim is to analyse the origin of the implementation of these production systems in agro-ecology or organic farming and to analyse biomass management practices on mixed crop-livestock farms involved in agro-ecology or organic farming (management of crop residues, fodder, effluents).

CHAPTER 1

I- Materials and method

This study was carried out in the Dano department in south-west Burkina Faso. It was carried out in 6 villages (Figure 1), namely Tambipkéré (coordinates), Dayéré , Tambiri, Yabogane, Complan, and Balembar (3°6'5 "W; 3°0'30 "W; 2°54'55 "W; 2°49'20 "W) and (11°1'W "N; ll°6'30 "N; ll°12'0 "N and 11°17'30 "N).

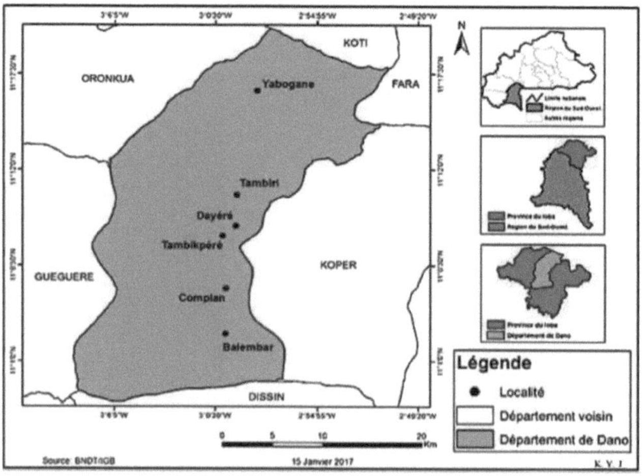

Figure 1: Location of study villages

To determine the different types of mixed crop-livestock farms involved in organic farming in south-west Burkina Faso, a qualitative approach was adopted. Surveys were carried out in these areas, where we submitted a questionnaire to a sample of 30 farm managers. The farmers surveyed were those involved in growing organic cotton, which relies heavily on the production of organic manure. The farms were divided into three types of mixed crop-livestock farming established by the *'Ecological Intensification and Design of Innovations in the Agro-Sylvo-Pastoral Systems of West Africa'* (DP ASAP) partnership research and teaching scheme (farmers, breeders and agro-breeders) for the cotton-growing areas of West Africa (Vall *et al*., 2006, 2012).

I-1 Identification of the types of agro-pastoral farms involved in organic farming

Analyses carried out using Excel software enabled us to distinguish types of agro-pastoral farms involved in organic farming on the basis of their structure. The variables selected for these analyses were : the total cultivated area (STC), the number of animals (tropical livestock unit or UBT), the number of draught oxen (BdT), the share of cotton, maize, sorghum and millet in the crop rotation, the number of farm workers, the distribution of fields and the distance of fields from concessions, farm equipment (farm implements and animal-drawn or motorised means of transport), farm families and workers, average production on the farms and their economic results from farming activity.

I-2 Study of biomass management practices

To analyse biomass management practices on the mixed crop-livestock farms identified, we used the systemic farm analysis approach proposed by Jouve (1992). This approach makes it possible to examine the agricultural practices implemented by the farm and the operating mode of the production system. We characterised the structure of the production systems (family, workforce, equipment, capital, land parcels, livestock composition), studied the operation of the production systems, the cropping system with crop rotation, crop rotation, crop succession, technical itinerary, management of soil fertility and crop residues, the livestock system with animal reproduction and management, sanitary practices, effluent management and animal feed).

CHAPTER 2

II - Results and discussion

Results

The results of this study include an analysis of the origins of the implementation of these production systems in organic farming, and an analysis of biomass management practices on mixed crop-livestock farms committed to organic farming (management of crop residues, fodder, effluents).

1- Diversity of agropastoral farms involved in organic farming

The first and most numerous types of farm practice agriculture. We distinguish between farmers according to the size of their cultivated area (4.8 ha). These farmers have at least limited animal traction equipment and a small herd (cattle, small ruminants, etc.). The average herd size is 5.4 LU. These farmers mainly produce cotton and cereals (millet, sorghum, maize, etc.) for sale and self-consumption on limited areas (4.8 ha). Livestock farmers have large herds of cattle (plough cattle (3.2%); breeding cattle (22.9%), small ruminants (41.8%), donkeys (2.2%) pigs (2.7%) and poultry (27.1%). The average tropical livestock unit (TLU) among livestock farmers is 27. Farmers cultivate an average of 6 ha of farmland. Like farmers, livestock farmers have the same size areas and produce the same crops. They are well equipped with animal traction and 50% are motorised. Finally, the agro-pastoralists in these areas cultivate large areas of land, averaging 7.6 ha. They grow cotton for marketing and cereals for their own consumption. They also own livestock with a tropical livestock unit of 22.2. Livestock comprises plough cattle (3.9%), breeding cattle (16.9%), small ruminants (34.5%), donkeys (1.3%), pigs (5%) and poultry (38.4%) (Table 1).

Table 1: Main types of farm

Variables	*Farmers*	*Farmers*	*Breeders*
Nb (individuals)	17	*8*	4

Age (year)	47,47	50,5	48
Agricultural area (ha)	4,8	7,6	6
Livestock (livestock units)	5,4	22,2	27
Family and worker	7,3	10,8	8,8

2- The origins of these production systems

a- History of the farms

The farms of farmers, agri-breeders and stockbreeders are characterised by the youth of their structure. In particular, 35% of the farms run by agri-breeders are over 25 years old, whereas those run by breeders and farmers appear to be older (25% of breeders' farms are over 36 years old and 5% are over 30 years old). However, overall, the farms are marked by their young age (2 to 25 years) **(Figure 2).**

The reasons why the sample settled are inheritance (after the death of the father), negotiated separation, and marriage for some. In this case (marriage), the woman uses the man's (the husband's) land for her settlement. In the case of inheritance, all assets are passed on in full, whereas in the case of separation, production assets (animals, land, equipment, assets) are shared between family members.

Figure 2: *Distribution of installation dates for farms in the sample, by farm type*

b- Reasons for organic farming

There are a number of reasons why farmers in south-west Burkina Faso have become aware of the need to move towards ecological intensification. Farmers in the area have seen a rise in land degradation and a reduction in arable land. The reasons for organic farming are the harmful effects of chemical inputs on the soil and the environment. Organic farming, which does not require sufficient expenditure (because it requires more credit to buy chemical inputs), has a positive long-term effect on the soil with organic fertiliser (3 years) compared with conventional farming (1 year). Some of the farmers said that organic farming is more profitable and more economical. From a health point of view, the use of herbicides, insecticides, etc. was causing health problems (body infections). Moreover, belonging to an organic farming movement aroused their interest in its production. And 100% of the sample surveyed belonged to a movement involved in ecological intensification (Reference: Dano organic farming database).

c- Advantages and constraints of organic farming

The advantages of organic farming highlighted by producers in the area are more related to soil productivity and the economics of their production. The advantages cited are good crop yields and long-lasting soil fertility (3 years), with the apology of a single spraying during this period (declarative title). On the economic front, the producers were enthusiastic about the price of organic produce. Organic cotton costs 325 FCFA per kilogram, compared with 225 FCFA for conventional cotton. Thanks to this, the farmers say they are able to meet their own expenses and family needs (children's schooling, etc.). However, despite the success of organic farming, there are a number of constraints to production. The disadvantages cited by producers are the difficulties of transporting organic manure: collection and transport are major constraints on the widespread use of straw. It is not easy, and certainly not profitable, to transport large quantities over long distances and quickly, at least as they are obtained. This justifies their widespread use locally at the production sites; stagnation of water in the fields during the winter destroys part of the plants, the long period of drought slows down the development of the plants, resulting in low yields at harvest, destruction of the plants by soil fauna (insects, ants, etc.).

3- Composition and structure of farms involved in organic farming

a- Agricultural area

In most cases, the largest cultivated areas are used to grow sorghum, cotton, maize and millet. On the other hand, the areas sown to rice and cowpea are very small. Although not as important, groundnuts and sesame are not neglected by the farm types (Figure 3).

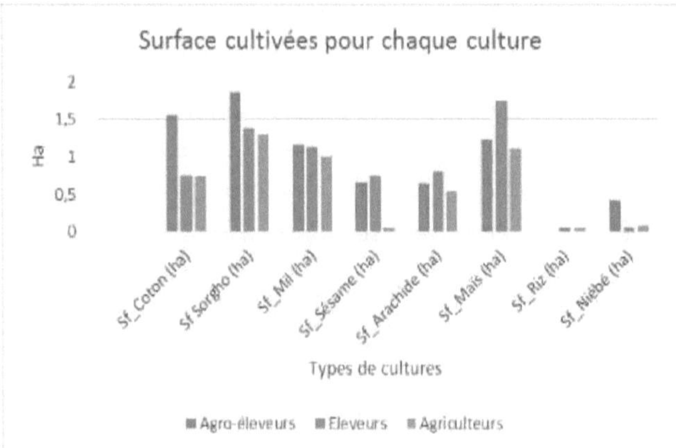

Figure 3: Area cultivated for each crop

These larger cultivated areas belong to agro-pastoralists and livestock farmers, especially for sorghum, cotton and maize (Table 2). The reason is that they are better equipped with animal traction and farm machinery **(Table 2**), which makes it easier for them to extend their farmland. Farmers who are less intensive overall cultivate medium-sized areas for their production because they have fewer means of production.

However, given the difficulty of accessing land titles, extension strategies seem difficult to maintain for the next generation. It should also be noted that all the producers in the sample we met own herds of animals, although these are smaller among the farmers.

Table 2: Agricultural land

Variable	*Total cultivated area (ha)*	*Distribution of fields*	*Distance from fields (km)*
Farmers	4,8 ± 1,8	Cotton, cereals	0.5 à 1.53 ± 1.43
Farmers	7,6 ± 3,5	Cotton, cereals	0.75 à 1.63 ± 0,97
Breeders	6 ± 1,8	Cotton, cereals	0.01 à 1.67 ± 0.53

It was found that crops sown on large areas are further from the concessions than small areas.

Agro-breeders, who are the most dominant, have farms that are further away, with an average distance of 0.75 to 1.63 km, compared with breeders and farmers whose farms are closer to the concessions, with average distances of 0.01 to 1.67 km and 0.5 to 1.53 km respectively.

The majority of farms have agricultural equipment, although not all farms have the same level of equipment (Table 3) and none of the farms surveyed have a seed drill. With regard to agricultural equipment, agro-pastoralists and livestock farmers have the most equipment, such as ploughs (1.4); (1.8), weeders (1); (1) ridgers (1.3); (1) and treatment equipment (0.8); (1) respectively. The least intensive farmers also have this equipment, but in smaller numbers than the other types of farmer, with an average of 0.8 ploughs, 0.6 weeders, 0.5 ridgers and 0.6 sprayers. During data collection, the producers all mentioned that they did not own a seed drill as part of their agricultural equipment.

As far as transport equipment is concerned, all the growers do not own tractors. However, they do have motorised transport. Some of the farmers do not own a tricycle, but do own a 0.5 motorbike on average. As well as motorbikes (1 and

0.6), the most motorised farmers and stockbreeders also have tricycles (0.1 and 0.5), which they use to transport their produce.

Table 3: Agricultural equipment.

Variable	*Farmers*	*Farmers*	*Breeders*
	Agricultural equipment		
Plough	0,8 ± 0,8	1,4 ± 0,7	1,8 ± 0,9
Seed drill	0	0	0
Weeder	0,6 ± 0,5	1 ± 0,5	1 ± 0,8
Bumper	0,5 ± 0,5	1,3 ± 0,7	1
Treatment equipment	0,6 ± 0,6	0,8 ± 0,5	1
	Transport equipment		
Trolley	0,4	1,3	1,5
Gondola	0,1	0	0
Tractors	0	0	0
Tricycle	0	0,1	0,5
Motorbike	0,5	1	0,6
Bike	2,3	4	2,3

b- Family and farm workers

On all farms, workers are essential to the smooth running of farming activities. They constitute the daily workforce on the farms.

The average number of workers on these farms is (10.8 ± 6.2); (7.3 ± 4.4) and (8 ± 1.15) respectively for agri-breeders, farmers and livestock breeders. Also during production, farm managers call on outside labourers two to three times a year to help with weeding, ridging and/or harvesting.

c- Agricultural production

Production yields are known at the end of the season. Growers start harvesting in

October and the harvest generally ends in December. Figure 3 shows the yields obtained in 2015 by the sample surveyed. As the graph shows, agri-breeders and livestock farmers are the producers who recorded the highest maize yields. Yields for all other crops were satisfactory and virtually the same for all producers.

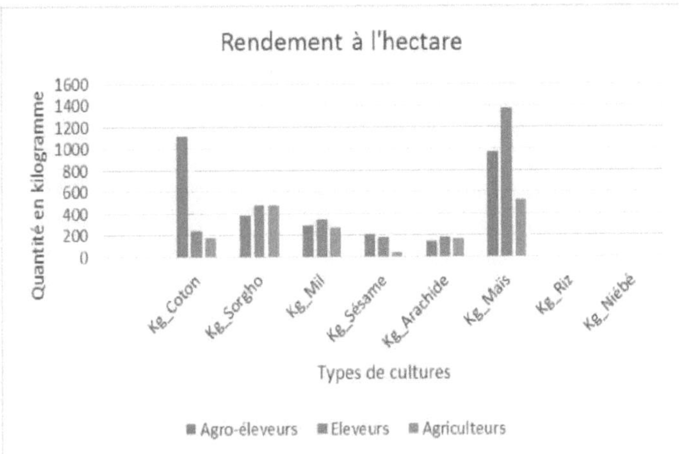

Figure 4: Production yield per hectare

At the end of the harvest, the products transported are intended either for sale or for self-consumption (Figure 4). However, if there is a shortage, agricultural products are bought. Of all these products, only cotton is fully sold at the end of its production and the other crops are practically consumed by the farmers and their families. The figure below shows the routes taken by agricultural products, in particular cotton, maize, millet, sesame, sorghum and groundnuts.

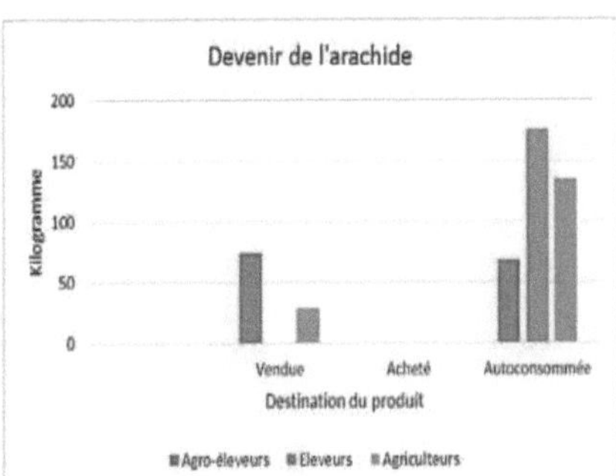

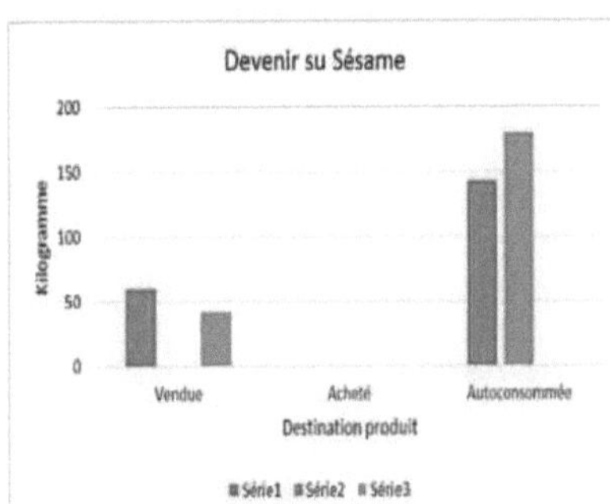

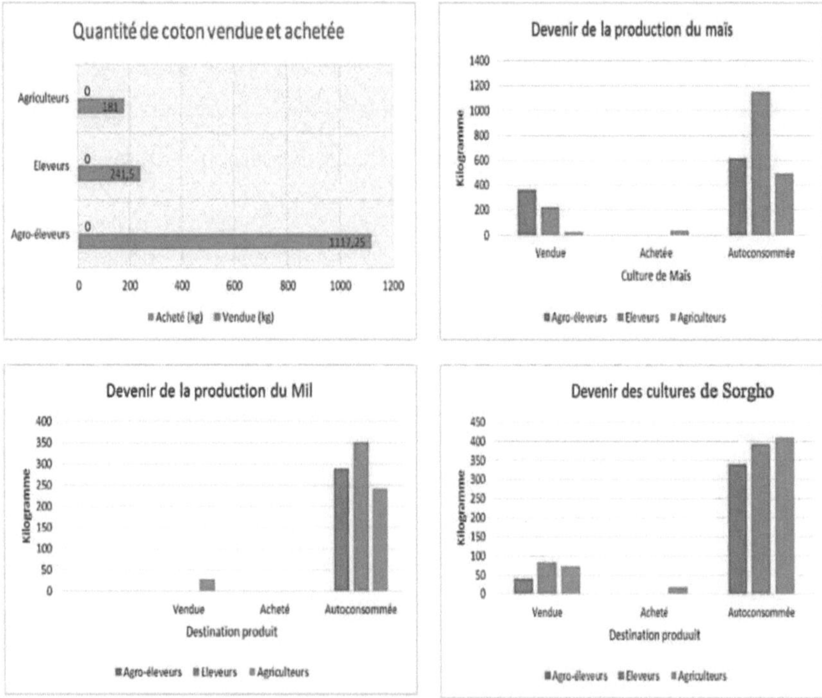

Figure 5: Destination of different agricultural products

3- Economic results of agricultural activity: the case of cotton and maize

In all cases, the economic results of agricultural activity are promising. At the end of the day, farmers' products are practically self-consumed. It is only the need for money that necessitates the sale of some products.

Table 4: Cotton

Variables	*Expenditure (FCFA)*	*Revenue (FCFA)*	*Balance sheet*
Farmers	32.150	156.150	Positive
Farmers	10.382,4	59.033,8	Positive
Breeders	8.250	78.562,5	Positive

Table 5: Maize

Variables	*Expenditure (FCFA)*	*Revenue (FCFA)*	*Balance sheet*
Farmers	8.937,5	51.250	Positive
Farmers	7.647,1	4.411,8	Positive
Breeders	15.000	27.500	Positive

A- ***Study of biomass management practices***

1- The growing system

a- Crop rotation

All the producers in the sample surveyed have an average of three fields. Some practise crop rotation and others fallow for short periods (2 years). On each plot, there is a succession of crops whose land use changes from one year to the next (rotation). For all the producers, the crops sown are practically the same. Livestock farmers and agro-pastoralists invest most in the production of maize, millet and sorghum. Farmers are limited to medium-sized areas but also produce the same crops. In addition to these main crops, producers grow sesame, rice, cowpeas and groundnuts on small areas. The sample excluded the use of mineral fertilisers, and all resorted to the production of organic manure to improve soil fertility, albeit in very different doses. They produce organic manure by making manure pits, heap compost and pit compost.

b- Technical itineraries and fertility management for cotton and maize crops

In order to achieve a good yield at the end of the harvest, there are certain production principles to be observed. Maize and cotton are among the crops most sown by the sample surveyed.

- Cotton

In the south-west zone of Burkina Faso, surveys of the sample indicate that agricultural work begins in March and ends after the harvest between November and December. Cotton seed is supplied by the UNPCB structure. Producers begin by emptying the organic manure and then spreading it collectively between March and May. Breeders are the producers who contribute the greatest quantity of organic manure per hectare, with an average of 3125 kg **(Figure 5).**

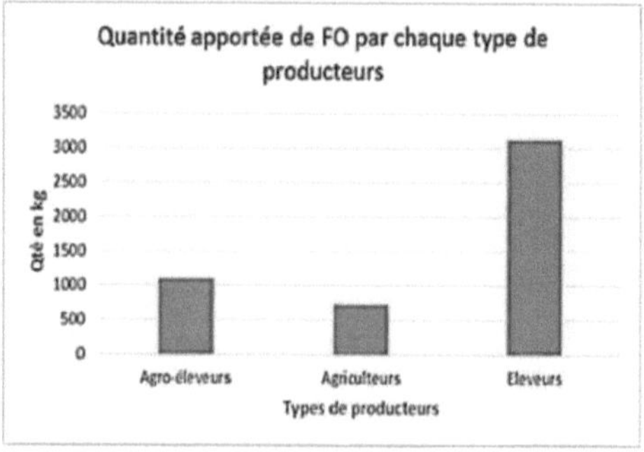

Figure 6: Quantity of organic manure applied to cotton crops

Farmers contribute an average of 1093.75 kg and farmers 712.744 kg. After spreading, the soil is worked either by ploughing or ridging. Farmers then resort to weeding and ridging. **Figure 6** shows how each type of farmer maintains their cotton and maize crops.

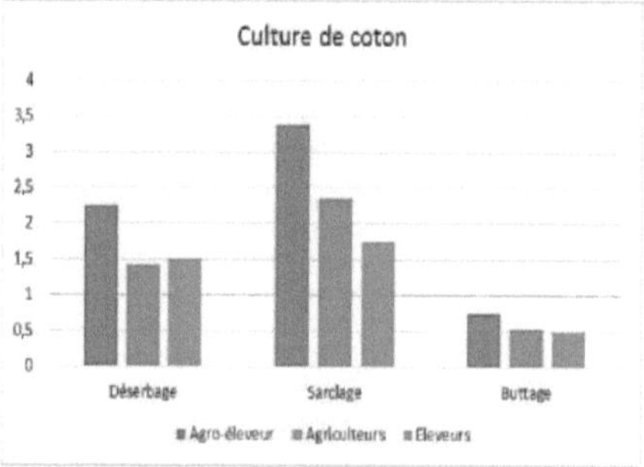

Figure 7: Proportion of different actions carried out on cotton fields

Farmers make the most passes for crop maintenance (weeding, weeding, ridging), followed by livestock farmers and then farmers. For weeding, the average numbers are 2.25, 1.5 and 1.41 respectively for agri-breeders, livestock farmers and farmers. Agro-pastoralists weeded cotton crops an average of 3.37

times, farmers an average of 2.35 times and pastoralists around 1.75 times. Ridging is carried out once a year by all types of producer.

- Corn

The technical itinerary for growing maize is no different from that for cotton. The phases and periods of production are similar. The emptying and spreading periods are practically the same. However, there are a few differences in each section. Figures 6 and 7 below show the quantities of organic manure applied to the maize field and the actions taken in the crop.

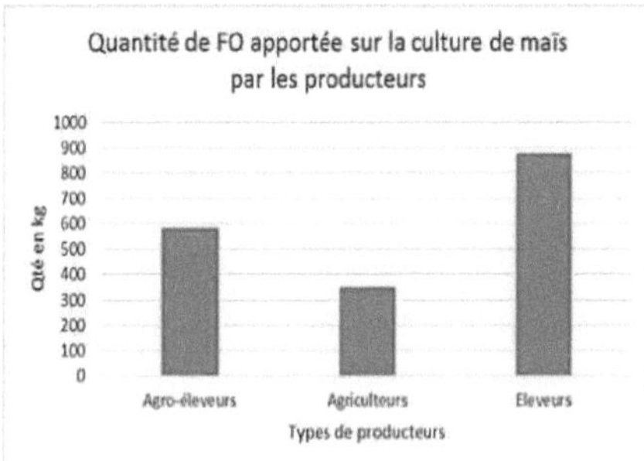

Figure 8: Quantity of organic manure applied to maize crops

The figure below shows that farmers optimise the use of organic manure on their farmland. The average amount of manure applied to maize fields by livestock farmers is 875 kg. Those for agro-pastoralists and farmers were 583.33 kg and 348.28 kg respectively. The figure below explains the actions taken by the various farms during maize production (Figure 9).

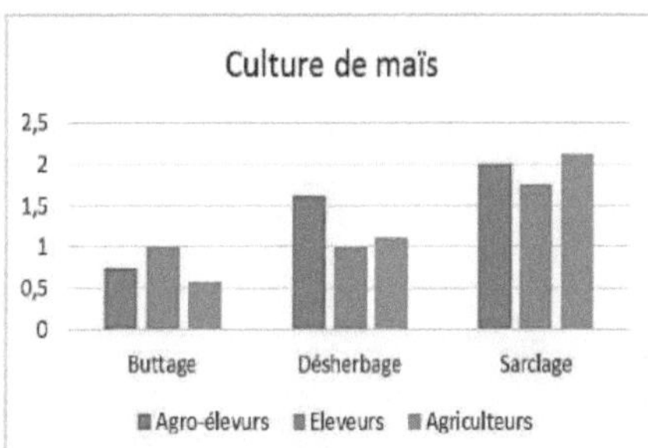

Figure 9: Proportion of different actions carried out on maize fields

c- Biomass management practices

To study the pattern of crop residue use, a principal component analysis (PCA) test was carried out (Figure 8). The principal component analysis was expressed as two axes that explained 62.19% of the total variability observed. 33.07% on the abscissa axis and 29.12% on the ordinate axis. This text allows us to understand the role played mainly by the different types of crop residue depending on the speculation.

Axis 1, which explained 33.07% of the total variability, was defined by two variables, the VP of the UP (kg) and fodder (kg). These parameters were negatively correlated with the axis (Figure 8b). In contrast, the residues were marked by the outside and burnt VP. Axis two, which represents 29.12% of total variability, was defined by two variables: residues used to produce organic manure, which were negatively correlated, and residues left in place, which were positively correlated.

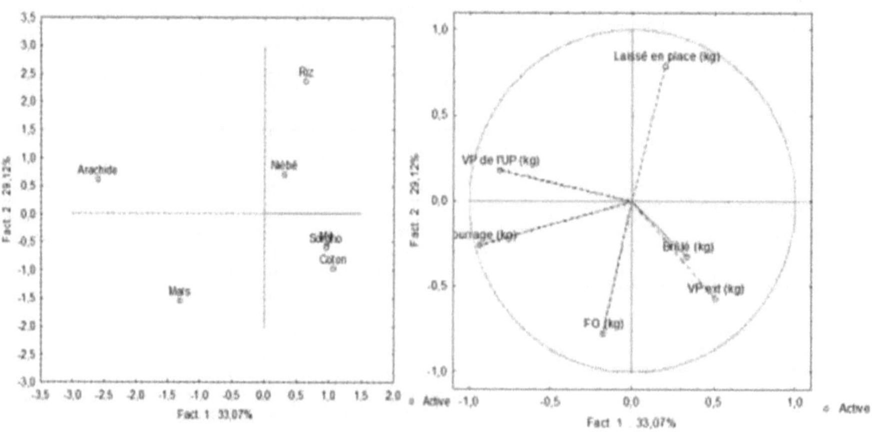

Figure11: a- des différentes types de cultures

b- le chemin des résidus de cultures

a- Biomass management practices

After the harvest, crop residue management is very important for the following season. Farmers use the residues to feed animals or produce organic manure

(Figure 9). However, some of the residues are left in the fields to be used for grazing, while others are stored for distribution during the hunger gap (dry season). Figure 9 above shows the different ways of managing residues from different crops. While cotton, groundnuts, sorghum and millet are mainly used for grazing or burning, and to a lesser extent for the production of organic manure (composting), cereal straws (especially maize) are mainly used for the production of organic manure (coarse straw or litter) and for animal feed, mainly by agro-pastoralists and livestock farmers who have greater needs for their herds. Legume tops are used entirely for animal feed. The figure below shows the average quantity of residues provided by each type of farm. It shows in detail the workload or motivation of the farms in managing biomass. In each graph, livestock farmers and agri-farmers store more residues, except in the case of maize, where farmers stored more maize residues.

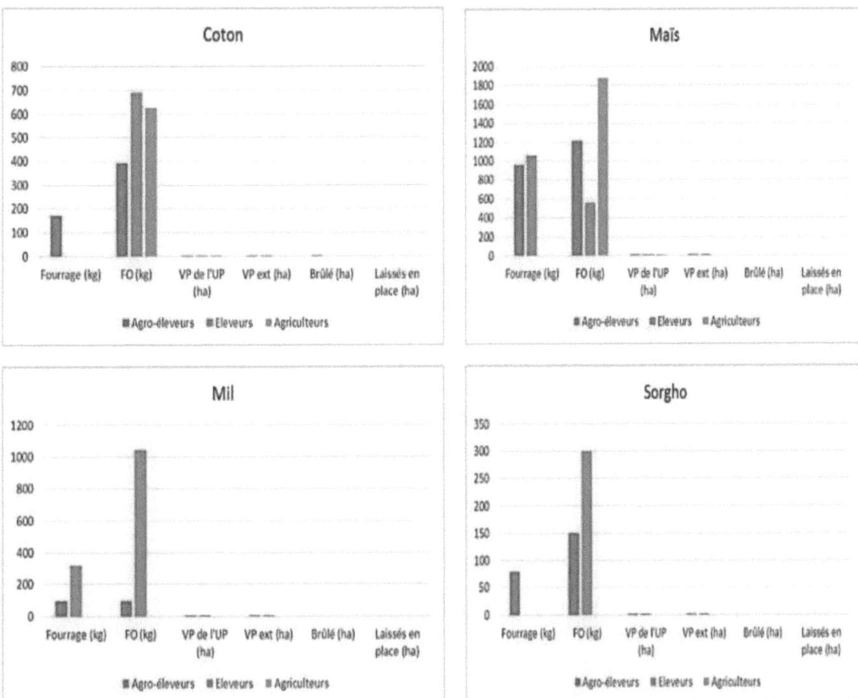

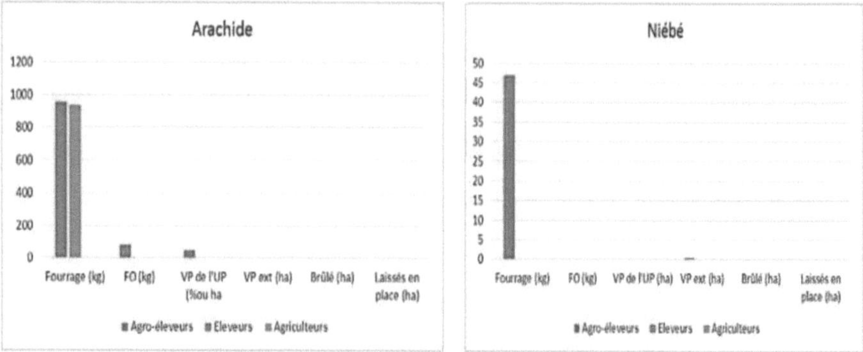

Figure 11: Farm management of crop residues

d- Organic manure production practices

Producers in Ioba province are characterised by the production of pit manure and compost in heaps. For all types of farmer, organic manure is produced in the plot or, for some, in the field.

They start by setting up a pit, which may be a simple hole or built with materials such as cement, stones and sand. They start by filling the pits with cereal straw, animal dung, household waste, ash, etc. Production of other types of manure is very limited.

The average age of the pits installed by agro-pastoralists is 8 years, 11 years for livestock farmers and 7.5 years for farmers. The production of compost in heaps is very recent.

Producers received training from UNPCB in 2016, and only a handful of producers have actually produced compost. Heap compost is a recent production of organic manure, so the activity of emptying it has not yet been carried out. The pits are watered at all times with household water throughout the period. The livestock farmers in the sample produce the most organic manure, with an average of 4,618.75 kg. Agro-pastoralists came second with an average quantity of 3593.75 kg, while less intensive farmers produced the lowest quantity, estimated at an average of 1856.36 kg (Figure 10).

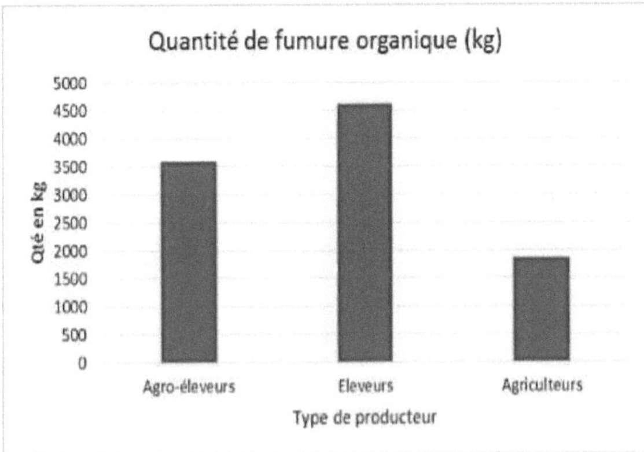

Figure 12: Average quantity of organic manure produced

*2- **Breeding system***

a- Characteristics of livestock production systems on surveyed farms

The livestock rearing practices of each sample surveyed are characterised by the composition of the herd of each type of producer and the movement within the herd. We distinguish between the production of cattle, donkeys, small ruminants, pigs and poultry (Table 6).

Table 6: Herd movements on different farms

Variables	*UBT*	*BoT (%)*	*Cattle (%)*	*Small ruminant (%)*	*Donkey (%)*	*Pork (%)*	*Poultry (%)*
Farmers	5,4	8,8	4,3	43,8	0,1	5,5	37,5
Farmers	22,2	3,9	16,9	34,5	1,3	5	38,4
Breeders	27	3,2	22,9	41,8	2,2	2,7	27,1

Draught oxen are used for harnessing and donkeys are used to transport farm inputs, crops, harvest residues and the FO. These animals play important roles on farms.

As far as inputs and outputs are concerned, this line of operation corresponds mainly to small ruminants. Animals such as large ruminants are inherited from the deceased head of the family. However, they are sold to be replaced after they are considered old and no longer have the necessary capacity to provide the work required.

Table 7: Herd movements on farms

Variable	*Farmers*			*Breeders*			*Farmers*		
	Nais (nb)	**Sale** (nb)	**Buy** (nb)	**Nais** (nb)	**Sale** (nb)	**Buy** (nb)	**Nais** (nb)	**Sale** (nb)	**Buy** (nb)
BdT	0	0	0	0	0,25	0	0	0	0
Cattle	2,37	0	0	5,75	0	0	0,41	0	0
éle									
Donkey	0,5	0	0	0,75	0	0,25	0	0	0.11
Goats	5,37	1,25	0	4	1	0	4	0,05	2,05
Sheep	6,25	0,625	0,75	6	0,75	0	1,58	0	0,17
Pigs	1,62	0,12	1,37	2.5	0,25	0	0,64	0,58	0,35
Poultry	5,25	14,62	0	6,25	26	0	3,70	5,70	1,58

b- Grazing animals

The time the animals spend grazing depends on the different periods and the availability of fodder.

During the cold dry season, grazing times are longer, because following the harvest season, there is a large availability of fodder, allowing the animals to graze better. However, in the hot season, the lack of feed limits the animals' grazing time.

Grazing time is short and the animals are fed from stocks kept for the lean season. In the winter months, grazing time is also limited. Farmers are busy working in the fields and some animals are used for ploughing (BdT) on the one hand and to prevent damage to farms on the other. During this period, animals could cause damage to other people's crops, which leads to conflicts between farmers. Depending on the weather, the animals all have the same grazing times. The animals are put in paddocks or on stakes at the same time for each type of producer, and the difference can be seen in the comparison between agro-pastoralists, stockbreeders and farmers. (**Figure 11).**

The figure shows that, whatever the period, agro-pastoralists and stockbreeders allow their animals to graze for a long time. As for the farmers, we have noted that they are more modest in these practices.

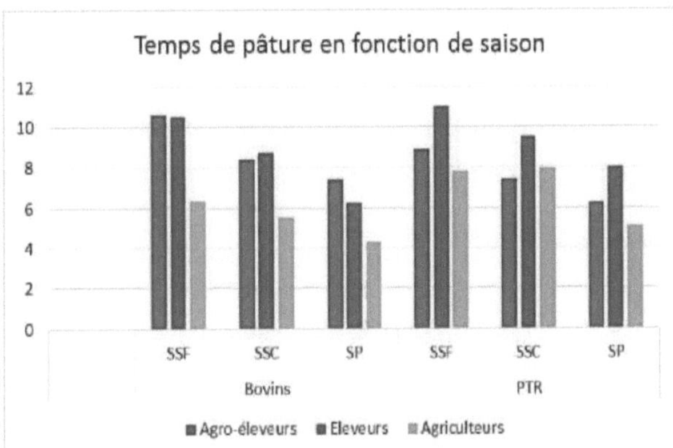

Figure 13: Different grazing times for the animals of each type of producer, depending on the three seasons.

c- Animal feed management

The producers in the Loba province sample store quantities of fodder according to the area cultivated with this product and the number of head of livestock available. The fodder most appreciated by the animals is groundnut stover, maize straw and rice straw. The other types of fodder (sorghum, millet, cowpea, etc.) are crop residues that are left in large quantities for animals to graze during the cold dry season. Farmers and breeders, who own large herds of animals, keep large quantities of fodder. However, they are practically equal in terms of the quantity of groundnut stover stored. (Figure 12).

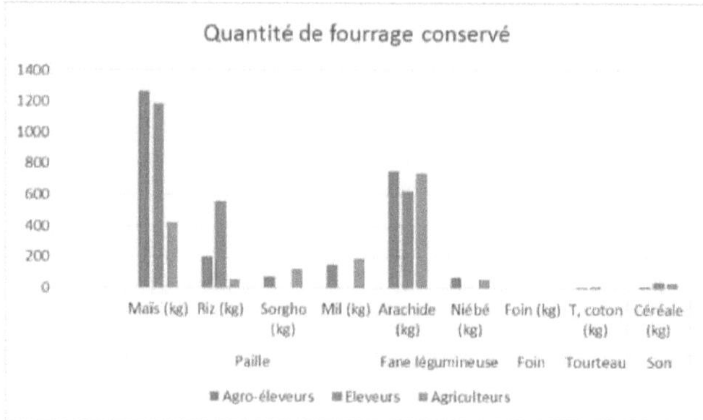

Figure 14: Quantity of forage stored

The various types of fodder stored are distributed to the animals during the dry and hot seasons. During this period, fodder is scarce and the animals roam around the concessions to signal their presence to the owner, while needing to feed on the stocks of fodder stored high up in sheds. Stockbreeders and agri-breeders stored large quantities of 2422.8 kg and 2548.1 kg respectively.

Table 8: Title

Variables	*Farmers*	*Farmers*	*Breeders*
Quantity stored (kg)	2548,1	1637,5	2422,8
Forage UBT(k g)	127,1	481	100,4
UBT feed (kg)	11,3	21,9	11

The average forage distributed by the producers in the sample was 121.1 kg of LU for agro-pastoralists, 100.4 kg of LU for livestock farmers and 481 kg of LU for farmers. UBT feed is that consumed by the animals. It was obtained by taking the difference between the total quantity of fodder and the UBT fodder.

d- Expenditure on the breeding system

Expenditure on animal production in the sample surveyed was limited to veterinary care and some animal purchases (table). As far as feed is concerned, the animals are fed on preserved fodder, and no commercial feed is purchased.

Variables	*Farmers*	*Farmers*	*Breeders*
Veterinary care of animals (amount) UBT	1761,4	6912,1	1652,8
Purchase of animals (amount) LU	721,2	7848,5	1070,9

At the end of each quarter, they are visited by the veterinary officer and prices are set according to the type of animal. The animals are either vaccinated or de-wormed (internal de-worming).

CHAPTER 3

III- Discussion

This study shows that the farmers in the sample are gradually moving towards ecological intensification. All practices linked to the use of chemical inputs are being forgotten by the farmers in the area. Depending on their individual objectives, each farmer has his own way of managing crop residues after harvest. The data collection and analysis work carried out on these farms has significantly improved our knowledge of biomass management practices and the sustainability of mixed crop-livestock farms in south-west Burkina Faso. The study also identified certain methodological limitations relating to the production of organic manure. Nonetheless, it offers some encouraging prospects for improving the analysis tool developed and for making better use of the results.

The diversity of agropastoral farms involved in organic farming

Initially, we characterised mixed crop-livestock farms through a typological study and described biomass management practices with 29 farm managers. This phase involved statistical analysis, immersion surveys and summary work. The different types of farms surveyed revealed the existence of three types of producer, distinct in terms of their structure and production systems. These three types of farm differ in particular in terms of livestock and farm area (total cultivated area). Thus, the components of a herd and farmland are the main source for determining farm types (Stéphanie, 2012). These surveys have shown that producers who intensify the most have a high yield per hectare. We compare this information with that of (Vall *et al.,* 2016; Blanchard *et al.* in press), whose studies on the family livestock project and the intensification path, respectively, showed that the farmers who intensified the most in terms of capital and inputs benefited from improved soil productivity, resulting in higher yields. In terms of soil fertility, farmers who diversified the most into cash crops had more fertile plots.

Study of biomass management practices on organic farms

The types of farmer are characterised according to their agricultural and livestock integration practices and their biomass management practices on their farms. The farmers in the sample surveyed have a biomass management or organic manure production method similar to the model described in the cotton-growing zone of southern Mali (Kanté, 2001; Blanchard 2013). As in the study by Tinguéri (2015), farmers encounter difficulties in transporting organic manure to their fields. These are farmers who are less intensive in their practices, some of whom are inclined to abandon everything in favour of other options. Leaving the country for neighbouring countries in search of a better life. In some cases, farmers are less intensive because of limitations in agricultural equipment. They plead for support from donors.

CHAPTER 4

Conclusion

Farms in south-west Burkina Faso are gradually moving towards ecological intensification. All practices related to the use of chemical fertilisers are on the way out. However, certain groups of producers are encountering difficulties in their biomass management practices, which is limiting their production and weakening their biomass management methods. Inadequate farm equipment is at the root of the failure to optimise biomass management. A policy needs to be put in place to encourage producers to invest more:

- ***Supporting producers in the acquisition of agricultural equipment,***
- ***Buying organic produce from producers at a more reasonable price, an alternative that would motivate them to invest more time in their work.***

This study should therefore be pursued through the following aspects:

- *To study the impact of waste on crops and the environment and the economic aspect of these practices and to assess the productive efficiency of these practices.*

CHAPTER 5

Reference

Bationo A., Kihara J., Vanlauwe B., 2007. Soil organic carbon dynamics, functions and management in West African agro-ecosystems. Agricultural Systems 94: 13-25p Blanchard M., Vayssieres J., Dugue P., Vall E., 2013. Local Tchnical Knowledge and Efficiency of Organic Fertilizer Production in South Mali: Diversity of Practices. Agroecol. and Sustain. Food Syst. 37 (6): 672-699.

Blanchard M., Fayama T., Yanga H., Dabire D., Kouadio K.P., Sodre E., in press Characterisation of intensification paths on mixed crop-livestock farms in western Burkina Faso and northern Côte d'Ivoire. 10p

DP ASAP, 2015. Base de données sur 250 exploitations agricoles des zones cotonnières du Burkina Faso, du Mali et de la Côte D'ivoire, Sous Access, INERA, CIRDES, IER, IDR, UPGC, CIRAD. Bobo Dioulasso (Burkina Faso)

FAO, 2014. Agricultural residues and agro-industrial by-products in West Africa ROM Italia 74 p.

Griffon M. 2009. Pour des agricultures écologiquement intensives : des territoires à haute valeur environnementale et de nouvelles politiques agricoles. Côte d'Amor, France: Editions de l'Aube and Conseil général.HOp

Jouve P. 1992 Approche systémique des modes d'exploitation agricole du milieu. Text extracted from the collective work "l'appui au producteurs : démarches, outils, domaines d'intervention" coordinated by M. Mercoiret CIRAD/SAR, published by the French Ministry for Cooperation and Development. 37p

Tingueri L.B., 2015. Understanding and characterising non-standard organic manure management practices in western Burkina Faso: Evaluating the sustainability of production systems implementing them. DEA dissertation, IDR,

Polytechnic University of Bobo-Dioulasso, 86 p.

Vall E, Chia E, Blanchard M, (2016) Partnership co-design of innovative farming systems. Cahiers Agricultures: (in press).

MIX
Papier aus verantwortungsvollen Quellen
Paper from responsible sources
FSC® C105338

Printed by Books on Demand GmbH, Norderstedt / Germany